PLANETS

CONTENTS

THE SOLAR SYSTEM

The Earth is one of nine planets that revolve around the sun. Millions of asteroids and comets also belong to this **solar system**. The sun is a star, far bigger than everything else in the solar system put together. Its powerful **gravitational pull** keeps the planets, comets, and asteroids in **orbit**. The sun is the source of light and heat for all the planets in the solar system.

Worlds Apart

The planets in our solar system are very different from one another. Some are made of rock, like the Earth, while others are nearly all gas. Pluto is extremely cold, while cloud-wrapped Venus is fiery hot. Some planets have rings and more than a dozen moons.

A STAR IS BORN

The sun—a medium-sized star—was formed about 4,600 million years ago. A vast cloud of gas, dust, and tiny specks of rock and ice began to shrink under its own gravitational pull. Most of the matter **condensed** at the center, where it swirled around and flattened out into a spinning disk. A sphere of denser gas grew at the center. This sphere shrank to become the sun.

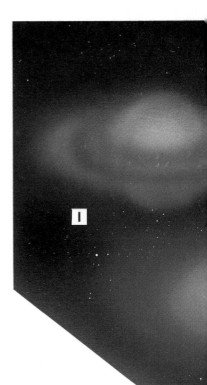

The Formation of the Solar System

1 The sun was formed at the center of a vast cloud of spinning gas, dust, rock, and ice.

2 At the same time, the **microscopic** particles of dust spinning around the new sun began to stick together, forming solid blocks, each several miles across.

3 The solid blocks condensed to form the planets. Four rocky worlds formed near the sun. Four icy worlds formed farther out. Their gravity pulled gas from the remains of the disk to build up into huge planets. Pluto, the outermost planet, is probably an icy block left over from the birth of the solar system.

Beta Pictoris
This **infrared** picture shows the formation of planets around a star called Beta Pictoris. The disk of gas and dust shows up here as red and yellow "wings."

2

3

MERCURY — THE SUN'S NEIGHBOR

Mercury is a small, barren world, pocked with craters and almost airless. As the nearest planet to the sun, it is very hot, although some permanently shaded craters at its poles are so cold they contain patches of ice. Astronomers think Mercury was originally a bigger planet, but as its core solidified, it shrank by two miles in diameter.

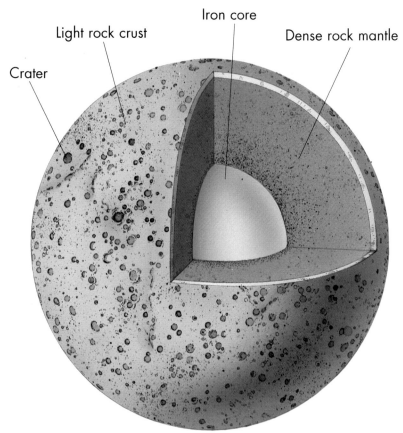

Iron core

Light rock crust

Dense rock mantle

Crater

Composition

The large core, made of iron, generates a **magnetic field**. The **mantle** consists of denser rock than the thin crust. The surface has a few small plains between the many craters.

The Long Day
Mercury orbits the sun quickly, but rotates at a much slower rate. This means that a Mercury day is longer than a Mercury year.

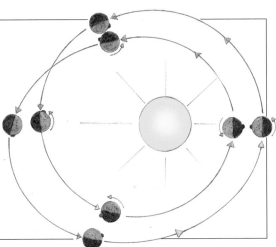

Iron Core
As Mercury cools down, its iron core gradually shrinks. The crust above has to buckle up, like the wrinkles on an old apple. These wrinkles show up as long narrow ridges that run round and between the craters. Some are more than 9,800 feet high and 310 miles long.

Mariner 10
The American probe, *Mariner 10*, is the only spacecraft that has visited Mercury. During 1974-75 it flew past Mercury three times, and sent back detailed pictures from its two cameras.

9

VENUS — VOLCANO PLANET

Venus and Earth are nearly the same size, but that's where the similarity ends. Venus is covered in clouds of concentrated sulfuric acid. Its "air"—made of unbreathable carbon dioxide—presses down with ninety times the pressure of Earth's atmosphere. The carbon dioxide traps the sun's heat and raises the temperature to 869°F, making Venus the hottest planet.

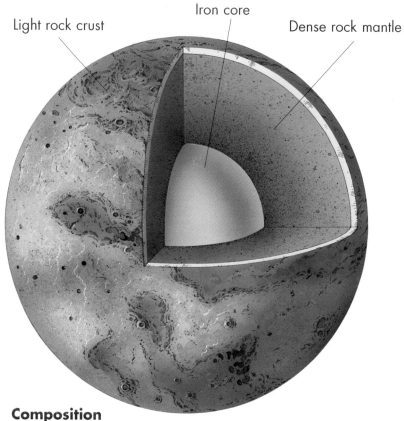

Light rock crust

Iron core

Dense rock mantle

Composition

Because Venus rotates very slowly, its iron core generates no detectable magnetism.

Hot molten rock rises from the mantle and erupts through the crust to form volcanoes.

Volcanoes

Venus is dotted with lava lakes and studded with up to 100,000 volcanoes, some almost as high as Mount Everest. Radar equipment that can "see through" Venus's clouds has revealed this volcanic crater, Sacajawea, which is 124 miles wide.

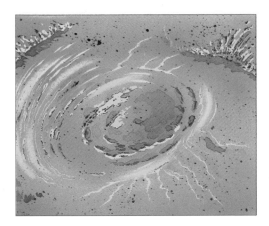

The Surface of Venus

The heavily protected Russian spacecraft *Venera 13* landed on Venus in 1982. Before succumbing to heat and pressure, it sent back **panoramic** views of a rocky volcanic slope under dull, orange clouds. Volcanic activity keeps most of the surface rocks covered with fresh lava.

EARTH—WORLD OF LIFE

The third planet from the sun—Earth—is different from every other planet in the solar system. From outer space, Earth has a blue tinge, because it is one of the few planets in the solar system with water. It is the only planet with much oxygen, and the only place in the solar system where life is found. The moon—which is one quarter the size of Earth—is dry, airless, and barren.

Earth
Our planet has just the right temperature for water to be liquid. As a result, life could begin in the oceans long ago.

Moon
The moon's weak gravity cannot hold on to any atmosphere. There is no life on the moon.

Continental Drift

Earth's continents are drifting. Over hundreds of millions of years, the world map has changed in many ways, from the shape of the oceans to the formation of mountain ranges.

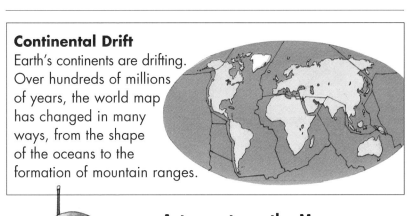

Astronauts on the Moon

Between 1969 and 1972, twelve American astronauts walked on the moon. They brought back a third of a ton of moon rocks and soil.

Origins of the Moon

The moon might have been created when another planet hurtled into Earth. The molten rock that sprayed from the collision cooled to form a ring of fragments that **coalesced** to form the moon.

13

MARS—THE RED MYSTERY

The reddish color of Mars inspired the Greeks and Romans to name this planet after their god of war. The length of a Martian day, the change of seasons, and the presence of polar ice caps are all similar to Earth's.

Recently scientists found possible evidence of simple life forms preserved for billions of years in a Martian **meteorite** that fell to Earth. This meteorite is the oldest of twelve that scientists have identified as Martian, and the only one in which possible signs of life have been found.

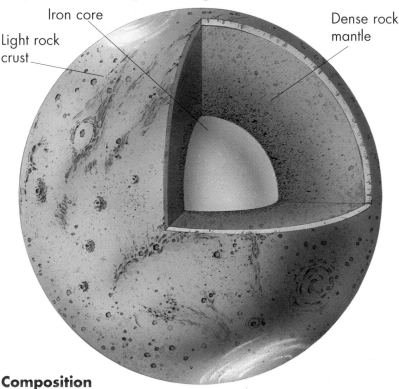

Iron core

Light rock crust

Dense rock mantle

Composition

Mars' small iron core produces no detectable magnetism. Molten rock has welled up from the mantle to form giant volcanoes and split the crust with deep canyons.

Olympus Mons

Although Mars is only half the diameter of Earth, it has some impressive features, including many **extinct** volcanoes. Olympus Mons is three times the height of Mount Everest and wide enough to cover Spain. It far overshadows the Earth's biggest volcano, Mauna Kea, on the Pacific island of Hawaii.

Olympus Mons

Mauna Kea

Valles Marineris

This canyon on Mars is as long as the continental United States is wide. The *Mariner 9* spacecraft first photographed it in 1971. Running into Valles Marineris are dried-up riverbeds, which indicate that early in its history Mars was much wetter and warmer than it is now.

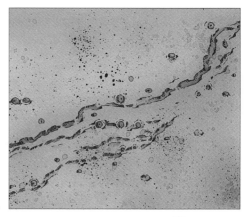

Life on Mars?

Long ago, when Mars was warmer, tiny cells may have formed, and then **hibernated** as the climate cooled. In 1976, two *Viking* spacecraft landed on Mars and collected samples of soil. Scientists added **nutrients** to the soil in an attempt to revive any cells, but this experiment failed.

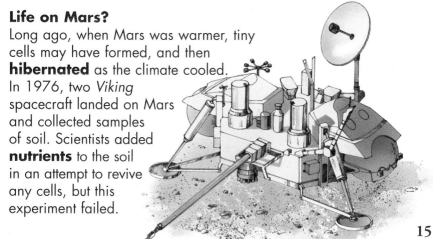

15

ASTEROIDS AND METEORITES

Between the orbits of Mars and Jupiter lies the asteroid belt. Asteroids are the solar system's building blocks, left over from the birth of the planets. The gravitational pull of nearby Jupiter prevented this space debris from assembling into new planets. Astronomers have discovered over 6,000 small rocky worlds here; the total number is probably much higher. The biggest of these asteroids is less than 620 miles across, and the smallest just a few miles in diameter.

Gaspra

From Earth, asteroids are so tiny that they look like points of light ("asteroid" means "looking like a star"). The first close-up view of one came in 1991, when the *Galileo* space probe, on its way to Jupiter, took a snapshot of Gaspra. Only ten miles across, this rocky world is shaped like a potato.

Meteorites

Small pieces of asteroids come in contact with Earth all the time. Those that fall to the ground are known as meteorites. Those that burn up in the atmosphere are called meteors.

Meteor Crater

Large chunks of asteroids can blast out craters when they hit the Earth. Meteor Crater in Arizona is .72 mile across. It was created by a million-ton lump of iron that crashed there 50,000 years ago. Older craters on Earth have largely been eroded away by rain, wind, and frost.

Killing Dinosaurs

The dinosaurs may have been wiped out by a large asteroid that hit Mexico 66 million years ago. The impact disrupted the climate and in turn affected plant life, which was the main source of food for many dinosaurs.

THE PROPERTIES OF THE PLANETS

The Sun

 Mercury

Venus

Earth

 Mar

Planet	Diameter at Equator (miles)	Average temperature (°Fahrenheit)	Period of rotation	Number of known moons	Planet
MERCURY	3,024	662	59 Earth days	0	MARS
VENUS	7,504	869	243 Earth days	0	JUPITER
EARTH	7,909	59	24 hours	1	SATURN

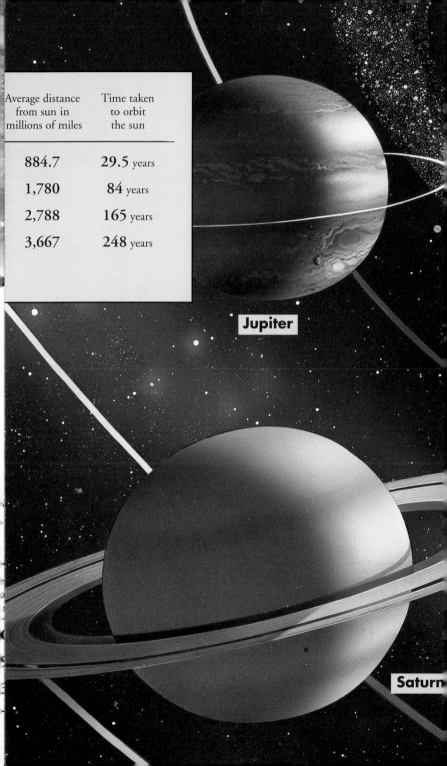

Average distance from sun in millions of miles	Time taken to orbit the sun
884.7	29.5 years
1,780	84 years
2,788	165 years
3,667	248 years

Jupiter

Saturn

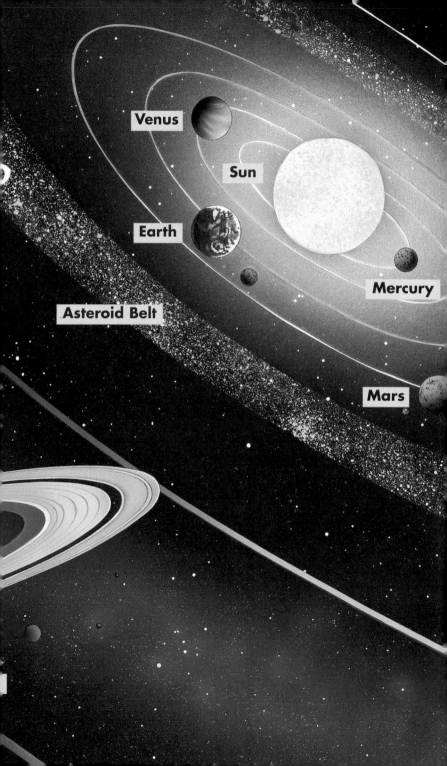

COMETS

A comet is a spectacular sight, with a long, bright tail stretching across the sky. A comet is made up of a head—or **nucleus**—of frozen gases and ice. The head is surrounded by an area of gas called a **coma**. As a comet approaches the center of the solar system, the sun's heat boils away the ice, creating the huge glowing tail.

Orbits
Comets follow long, looping orbits, often tilted up compared to the orbits of the planets.

Orbit of a comet

Orbit of Jupiter

Sun

Orbit of Earth

Anatomy of a Comet
A comet is made of four parts: gas and dust around a solid nucleus, a gas tail, and a dust tail.

Giotto Spacecraft
In 1986, the *Giotto* spacecraft sped through Halley's comet and photographed jets of gas spouting from the black surface of its nucleus.

25

JUPITER — GIANT PLANET OF GAS

Jupiter is the king of the planets. It is so big that all the other planets in the solar system could fit inside, and its gravity controls sixteen moons. This "gas giant" has no solid surface. It is made almost entirely of hydrogen gas, becoming denser and hotter toward the center. Among the bands of clouds in Jupiter's atmosphere is the Great Red Spot, a storm twice the size of Earth.

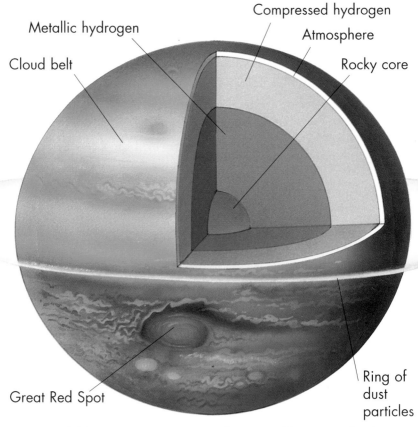

Compressed hydrogen

Metallic hydrogen

Atmosphere

Cloud belt

Rocky core

Great Red Spot

Ring of dust particles

Composition
Surrounding a core of molten rock, hydrogen is so compressed that it conducts electricity, like a metal. Above the sea of compressed hydrogen is a thin **methane** atmosphere.

Shoemaker-Levy 9
This comet broke apart and smashed into Jupiter in 1994. Resulting explosions left the planet with patches of black clouds the size of Earth.

Moons

Four of Jupiter's moons are bigger than Pluto, including the largest moon in the solar system, Ganymede. It is covered by a network of cracks. Dark Callisto is peppered with craters. A layer of ice on Europa may conceal oceans where sea life could exist.

Io

Io is covered with volcanoes that erupt yellow and red sulfur **compounds**. The spacecraft *Galileo* is monitoring these eruptions as it orbits Jupiter at a safe distance. *Galileo* has also dropped a probe into Jupiter's stormy atmosphere.

SATURN — WORLD WITH RINGS

Saturn, with its spectacular rings, is second in size only to Jupiter. This gas giant is so low in **density** that it would float in water—if there were an ocean big enough. Saturn's rings are so wide that they would stretch almost from Earth to the moon, yet they are less than a mile thick. Saturn also has more moons than any other planet—at least twenty.

Weather Systems

Saturn's cloud patterns are not as bright as those on other planets, but once every thirty years or so a huge storm breaks through, giving this planet a Great White Spot.

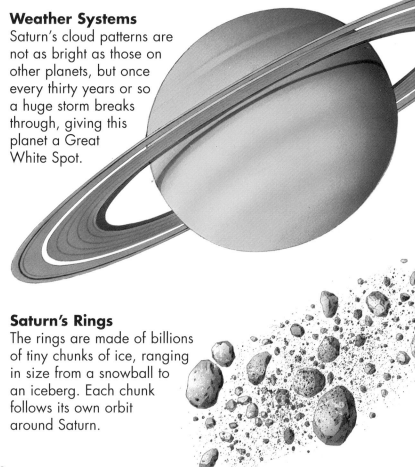

Saturn's Rings

The rings are made of billions of tiny chunks of ice, ranging in size from a snowball to an iceberg. Each chunk follows its own orbit around Saturn.

Mimas

Long ago, an asteroid or comet blasted out a huge crater on the surface of Mimas, leaving this moon with its eyeball shape.

Tethys

The surface of Tethys is cracked and cratered. Like Saturn's other moons, Tethys contains very little rock — it is made almost entirely of ice.

Titan

The largest of Saturn's moons, Titan is the only moon in the solar system with a thick atmosphere, denser than Earth's air. Titan is shrouded with orange clouds made of tiny **organic** droplets. This moon's surface may have lakes filled with liquefied natural gas.

Voyager 1

Voyager 1 sent back the first detailed pictures of Saturn as it sped past in 1980. It found several small moons and discovered that the rings are split into millions of narrow "ringlets."

An Earth in Deep Freeze

Conditions on Titan are similar to those on Earth shortly after our planet was formed. On early Earth, rain washed organic droplets into the oceans, where the first life began. Titan is too cold for rain, so its **"primeval soup"** has been preserved.

URANUS — PLANET ON A POLAR TILT

Uranus is so far away from Earth and so faint that it can only be seen through a telescope. This planet is tipped up on its side, probably because it was knocked over long ago by a smaller, wayward planet. Sometimes its north pole points toward the sun, and sometimes its south pole does. The seasons on Uranus are very extreme.

Water

Atmosphere

Composition

Uranus has no solid surface, but is water-rich. In its center is a core of molten rock. The deep blue-green atmosphere, made up mainly of hydrogen and methane, has few clouds.

Rocky core

Rings

Uranus has a set of very narrow, dark rings, tipped up like the planet itself. Astronomers only had hints of these rings until 1986, when the space probe *Voyager 2* flew past Uranus and sent back detailed pictures. The rings are made of tiny blackened pieces of rock and ice, as dark as coal, which orbit Uranus like miniature moons.

Rings

Each of Uranus's fifteen moons is made of a mixture of rock and ice. The five biggest are visible with a powerful telescope, and *Voyager 2* discovered ten more tiny moons. While most moons are named after characters from Greek mythology, those of Uranus bear names from English literature, especially the plays of Shakespeare. Titania, Oberon, Miranda, and Puck are among them.

Ariel
Huge cracks run around Ariel's equator between big, dark craters. Ariel's cracks may have resulted when water inside the moon froze and expanded.

Umbriel
A thin, black layer coats the icy exterior of Umbriel, darkest of the moons. It has only one bright crater, called Wunda.

Miranda
A patchwork world of enormous cliffs and tracklike markings, Miranda may have been broken up and then reassembled.

William Herschel
Herschel was a German astronomer who built telescopes and studied the stars. In 1781, he discovered a previously unknown planet. It was named Uranus, after Saturn's father in mythology.

NEPTUNE — WORLD OF WIND AND WATER

Neptune, which is nearly the same size as Uranus, was discovered in 1846 by astronomers trying to determine what was pulling Uranus off course. It was named after the Roman god of the oceans — which is appropriate, as we now know that Neptune is a watery planet. Most of our knowledge of Neptune has come from *Voyager 2*, which swept past the planet in 1989.

Composition

The interior of Neptune is much like Uranus, but its atmosphere is much cloudier. It too is surrounded by a set of dark rings.

Direction of winds

Hurricane Winds

Neptune has the strongest winds in the solar system. *Voyager 2* discovered that this planet's Great Dark Spot is an enormous storm, bigger than Earth. In 1995, the Hubble space telescope showed that the Great Dark Spot had vanished!

Eight moons orbit Neptune. Triton is as big as the planet Pluto. It orbits Neptune in the opposite direction of the planet's own rotation—the only large moon in the solar system to do this. Little Nereid follows an elongated orbit around Neptune, and can be seen from Earth through large telescopes. The other six small moons were discovered by *Voyager 2.*

Triton
Triton is the coldest world in the solar system. It has a very **tenuous** atmosphere, made up largely of nitrogen. Frozen methane ("natural gas") forms white, icy caps at its poles.

Sooty Geysers
In 1989, astonomers found that dozens of **geysers** on Triton were constantly erupting. Nitrogen gas blasted dark soot high into the atmosphere. The wind blew the soot for hundreds of miles, creating dark streaks across Triton's surface.

Voyager 2
Between 1979 and 1989, this spacecraft visited four different planets: Jupiter, Saturn, Uranus, and Neptune.

33

PLUTO — THE FROZEN FRONTIER

Tiny, frozen Pluto is the smallest of the planets; it was found as recently as 1930. Although Pluto is usually the most distant planet, its elongated orbit crosses Neptune's path, and sometimes (as in the years 1979-99) it is closer to the sun than Neptune. Pluto's moon is fully half the planet's own size. Many astronomers regard Pluto and its moon as a "double planet."

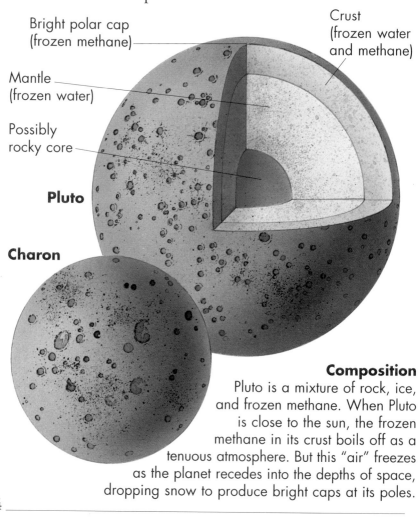

Bright polar cap
(frozen methane)

Crust
(frozen water
and methane)

Mantle
(frozen water)

Possibly
rocky core

Pluto

Charon

Composition
Pluto is a mixture of rock, ice, and frozen methane. When Pluto is close to the sun, the frozen methane in its crust boils off as a tenuous atmosphere. But this "air" freezes as the planet recedes into the depths of space, dropping snow to produce bright caps at its poles.

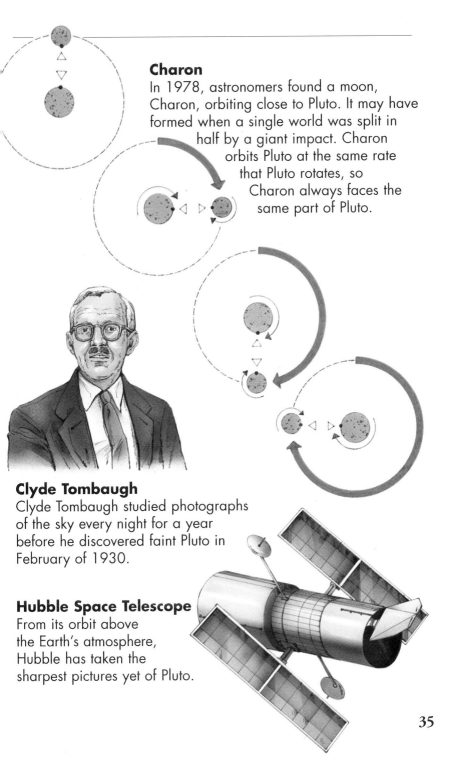

Charon

In 1978, astronomers found a moon, Charon, orbiting close to Pluto. It may have formed when a single world was split in half by a giant impact. Charon orbits Pluto at the same rate that Pluto rotates, so Charon always faces the same part of Pluto.

Clyde Tombaugh

Clyde Tombaugh studied photographs of the sky every night for a year before he discovered faint Pluto in February of 1930.

Hubble Space Telescope

From its orbit above the Earth's atmosphere, Hubble has taken the sharpest pictures yet of Pluto.

AMAZING PLANET FACTS

- **Planet or Not?** Some scientists contend that tiny Pluto is not a planet, but one of thousands of "ice dwarfs" orbiting the sun beyond Neptune. Most experts, however, still consider it to be a planet.

- **Fast and Slow** Mercury orbits the sun 1,028 times in the time it takes Pluto to complete just one orbit.

- **Back to Front** Although Earth — and most of the planets — rotate from west to east, Venus spins east to west.

- **Hottest World** Even though it's not the closest to the sun, Venus is the hottest world, with a temperature of 869°F. The coldest is Neptune's moon, Triton, with a temperature of -391°F.

- **Magnetic Fields** A magnetic compass needle wouldn't be much use on other planets. Venus and Mars have no magnetism, and on several of the other planets the magnetic field is opposite to Earth's, so the "north" end of the needle would point south.

- **Fastest Winds** Neptune has the fastest winds in the solar system, blowing at 1,240 miles per hour — ten times the force of a hurricane on Earth.

- **Annual Shrinking** Jupiter is shrinking, at the rate of one millimeter per year. This compression heats up the interior of the giant planet, generating infrared radiation that astronomers can measure from Earth.

- **Biggest Planet** Jupiter is the biggest planet, but it spins more rapidly than any other, with a "day" — from sunrise to sunrise — that is less than ten hours long.

- **Watch for Falling Rock** Many meteors or "shooting stars" seen at one time are known as a meteor shower. Most meteor showers are thought to be the debris of comets.

GLOSSARY

Coalesce To merge together to form a new object.

Coma An area of gas surrounding the nucleus of a comet.

Compound A substance formed from two or more separate elements: water is a compound of hydrogen and oxygen.

Condense To reduce a collection of matter to a more compact form.

Density The extent to which matter is packed together.

Extinct Something that has died out and no longer exists.

Geyser A spring that shoots up jets of hot water and steam.

Gravitational pull The force exerted by any large body that attracts things toward it.

Hibernate To become inactive.

Infrared The heat energy that is given out by atoms.

Magnetic field The area in which a magnet is able to affect anything made of iron.

Mantle The part of a planet between the crust and the core.

Meteorite A piece of an asteroid that has fallen to Earth's surface. A piece of an asteroid that burns up in the atmosphere before it reaches Earth is called a meteor.

Methane A colorless natural gas that can be found in marshes and swamps on Earth.

Microscopic Too small to be seen with the naked eye.

Nucleus, nuclei The center or centers of an object or objects.

Nutrient Nourishing food.

Orbit The path a planetary object follows as it revolves around another.

Organic Any substance made up largely of carbon.

Panoramic A complete view of the landscape.

Primeval soup Organic substances dissolved in oceans from which early life may have sprung.

Solar system The sun and the planets, asteroids, and comets that move around it.

Tenuous A thin consistency.

INDEX *(Entries in **bold** refer to an illustration)*